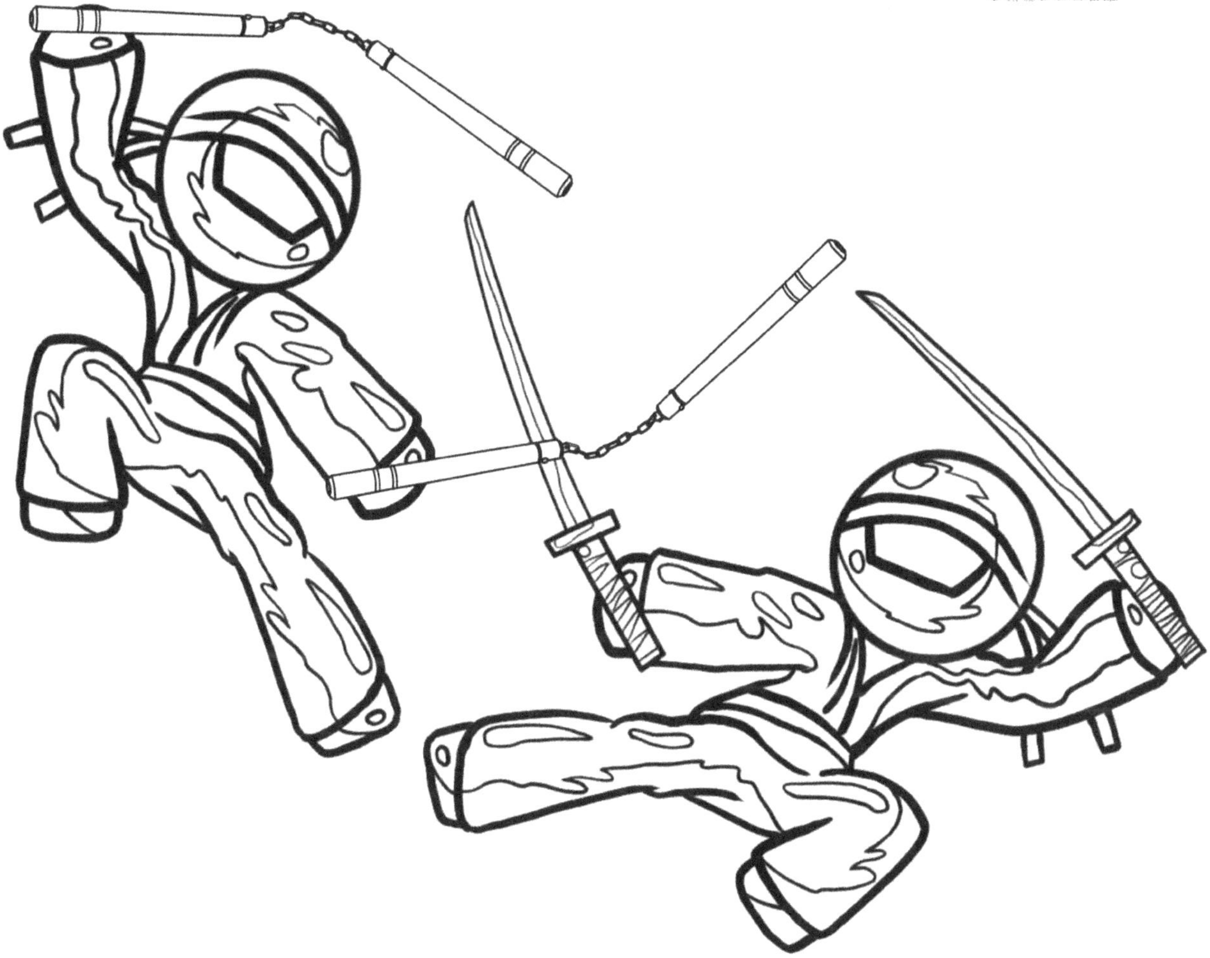

NINJA
MATH

≡ 1 or 2

How many Ninj'a Jumping ?
Circle the number you think is
the right answer.

= 2 or 3

How many Ninja's Jumping? Circle the number you think is the right answer.

≡ 3 or 4

How many Ninja's Jumping? Circle the number you think is the right answer.

page 3

☰ 3 or 4

How many Ninja's Jumping ? Circle the number you think is the right answer.

page 4

4 or 5

How many Ninja's Jumping ? Circle the number you think is the right answer.

= 5 or 6

How many Ninja's Jumping?
Circle the number you think is
the right answer.

= 6 or 7

How many Ninja's Jumping? Circle the number you think is the right answer.

≡ **7 or 8**

How many Ninja's Jumping ? Circle the number you think is the right answer.

= 8 or 9

How many Ninja's Jumping ? Circle the number you think is the right answer.

≡ 9 or 10
How many Ninja's Jumping? Circle the number you think is the right answer.

10 or 11

How many Ninja's Jumping?
Circle the number you think is the right answer.

11 or 12

How many Ninja's Jumping?
Circle the number you think is
the right answer.

=12 or 13

How many Ninja's Jumping? Circle the number you think is the right answer.

= 13 or 14

How many Ninja's Jumping? Circle the number you think is the right answer.

14 or 15

How many Ninja's Jumping?
Circle the number you think is
the right answer.

≡ 15 or 16

How many Ninja's Jumping? Circle the number you think is the right answer.

16 or 17

How many Ninja's Jumping ? Circle the number you think is the right answer.

≡ 17 or 18

How many Ninja's Jumping ?
Circle the number you think is
the right answer.

page 18

18 or 19

How many Ninja's Jumping ?
Circle the number you think is
the right answer.

page 19

≡ 19 or 20

How many Ninja's Jumping ? Circle the number you think is the right answer.

page 20

20 or 21
How many Ninja's Jumping ? Circle the number you think is the right answer.